This book belongs to:

_____

These are... Numbers

Written by
Cintia Roman-Garbelotto
Valentina Garbelotto
Illustrated by
Oana Voitovici

To my Mom and Dad, hence... Abu Tita and Abu Omar, for always believing in me.

This is number

One

Do you want to have some fun?

This is number Two

Can I count with you?

1, 2.

This is number Three

Please count with me 1,2,3.

This is number four

Let's count some more!

Oh, yeah!

One Two

3

Three

4

Four

This is number five

Can you jump

very high?

1,2,3,4,5.

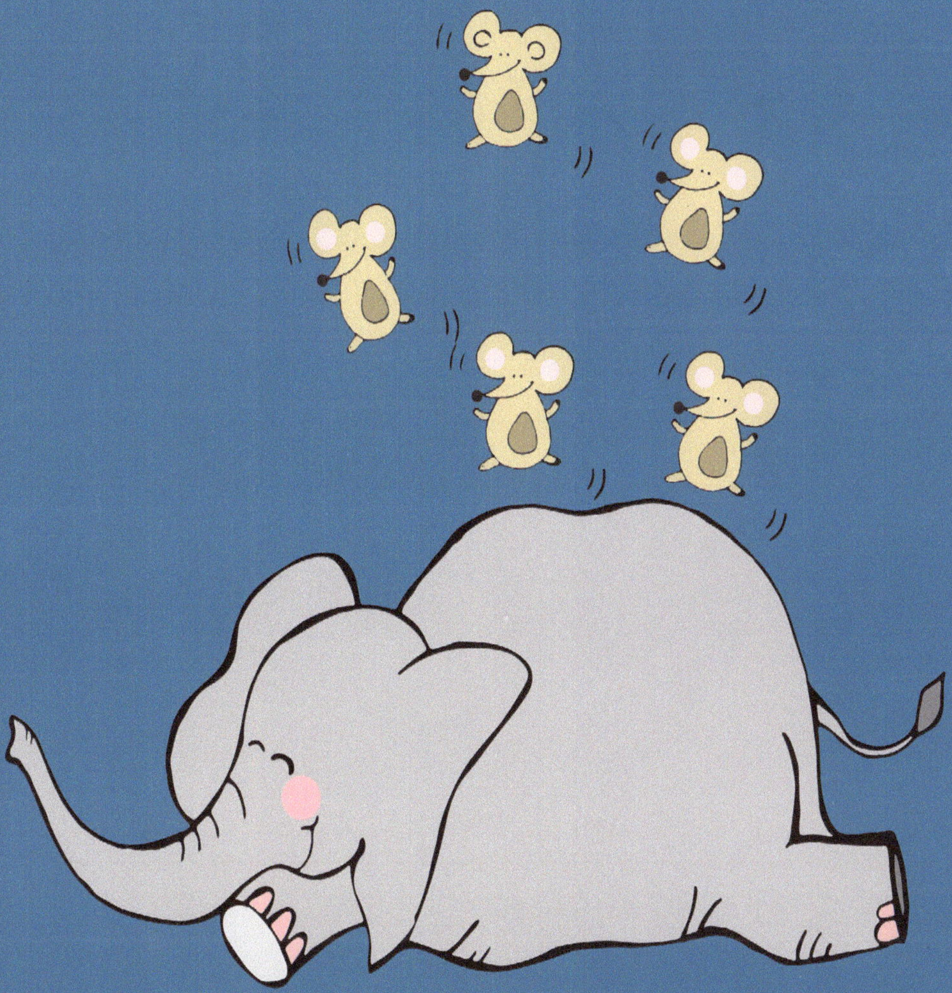

Making magic tricks!

1,2,3,4,5,6.

Let's go on
counting together!

1,2,3,4,5,6,7.

This is number eight

Don't stay up too late!

1, 2, 3, 4, 5, 6, 7, 8.

This is number
nine

Ready for
another rhyme?
1,2,3,4,5,6,7,8,9.

This is number Ten

10

Let's count again!

Here we go...

...again!

...and again!

1 One

Two 2

3 Three

4 Four

5 Five

6 Six

Seven 7

Eight 8

Nine 9

Ten 10

Can you find the numbers in the fish bowl?

www.ingramcontent.com/pod-product-compliance
Lightning Source LLC
Chambersburg PA
CBHW041615180526
45159CB00002BC/861